On a
Butterfly's Wing

"This exceptional book is an inspiration, a personal guide, and a call to action. A beautifully illustrated autobiography of a butterfly based on a true story, it spins a magic that captures hearts and leaves us with hope and a determination to do better. For ourselves, for nature – and for the butterflies. A brilliant read for all ages and a philanthropic tool for rewilding butterflies."

—Kris Tompkins, U.N. Environment Patron of Protected Areas, Tompkins Conservation

"Every so often a book comes along with a sweep and magic that means that you'll rarely find a copy languishing in a second hand bookshop because it has metamorphosed into a family treasure."

—Sir Tim Smit, founder of the Eden Project

"With a very original approach and beautiful aesthetics, On a Butterfly's Wing explains a very sensitive, very deep and, moreover, very well documented scientific story. The Painted Lady butterfly will never cease to amaze me, this time too, in the form of this little book, which is a real treasure."

—Constantí Stefanescu, European Butterfly Monitoring Scheme, Natural Sciences Museum, Granollers, Spain

On a Butterfly's Wing

Lessons from Nature on Embracing Change

Astrid Vargas
Award-Winning Conservation Biologist

To Amatxi and her clan

Contents

Prologue 6

I The Magic of Transformation
1. A Quick Overflight 10
2. Shedding and Growing 12
3. The Imaginal Soup 18
4. A Fateful Accident 26

II My Life as a Flightless Butterfly
5. Mission Amatxi 36
6. The Gregarious Globetrotters 48
7. Never Give Up 60

III Lessons from Nature
8. The Eternal Cycle 74
9. Looking Back 86

IV Regeneration
10. Attention, Here and Now 102
11. Looking Forward 106
12. Stardust 116

In Memoriam 118
Disappearing Butterflies 120
Further Flights 122
About the Author 124
Gratitude 126

Prologue

In our era of global change we need stories that offer hope and positive solutions to the challenges we face. *On a Butterfly's Wing* is such a story. It is a book written for the butterfly living inside of us all, waiting in its chrysalis form to grow wings and take flight in a new aspect of life.

On a Butterfly's Wing is a celebration of life and a tribute to our constant process of transformation. It is also a call for action for the protection of butterflies, offering positive suggestions on how we can team up with pollinators to help each other create a better world for both.

Fusing science, art, philosophy and conservation, this is the story of La Reme – a flightless butterfly who narrates her life experience from egg to death and beyond. Interwoven with La Reme's life cycle is the life cycle of Amatxi, a lovely lady who is facing her own death. Both stories are true, and they intertwine in reflections on our own transformations and on the unique privilege of being alive.

Throughout the book, La Reme speaks in the first person, and that is perhaps the only fictional aspect of this story. The rest, La Reme herself, her accident, Amatxi and her clan, the worldwide decline of butterflies, our need of mutual support…

All is real.

I
The Magic of Transformation

1
A Quick Overflight

They call me La Reme. I am a one-winged butterfly with a zest for adventure.

I spent my childhood as a caterpillar in Amsterdam and took my first flight inside an aeroplane, my own wings still folded inside my chrysalis.

I first spread my wings in Spain, where an unfortunate accident rendered me flightless even before I could use them. Yet, I managed to fly back to Amsterdam as a stowaway in another plane.

Despite my condition, I was lucky enough to find a mate back home and complete my life cycle as an adult, grounded butterfly.

Now, let me tell you my story in a few wingbeats...

Every flight begins with a wingbeat...

2
Shedding and Growing

I was raised in Amsterdam as part of a special butterfly taskforce, recruited to bring joy to the hearts of a close-knit human family.

Amatxi, their beloved matriarch, was very ill. Everyone knew her life was coming to an end, and the family wanted to do something to show her their love. Mother's Day was approaching and Amatxi's children and grandchildren decided to plan a special celebration, a tribute to life and its many transformations.

And so they recruited us, Painted Ladies, to help them out.

Timing was of the utmost importance, since all of us – the 56 members of Amatxi's Butterfly Taskforce – had to be ready to fly by Mother's Day, in early May.

I had just hatched from a tiny egg and eaten the shell, as we caterpillars do to get a head start in life. With a full stomach, I looked around me and realized that there were lots of other caterpillars, my siblings, and plenty of food for us to eat.

A butterfly of the thistles

Without much ado, I built a silk shelter to protect myself from predators and began to eat the tasty leaves of a thistle.

I love thistles, as all butterflies that look like me do. That's why humans call us *Vanessa cardui*, which means *Butterfly of the thistles*. But we are more commonly known as Painted Ladies, which sounds much nicer, don't you think?

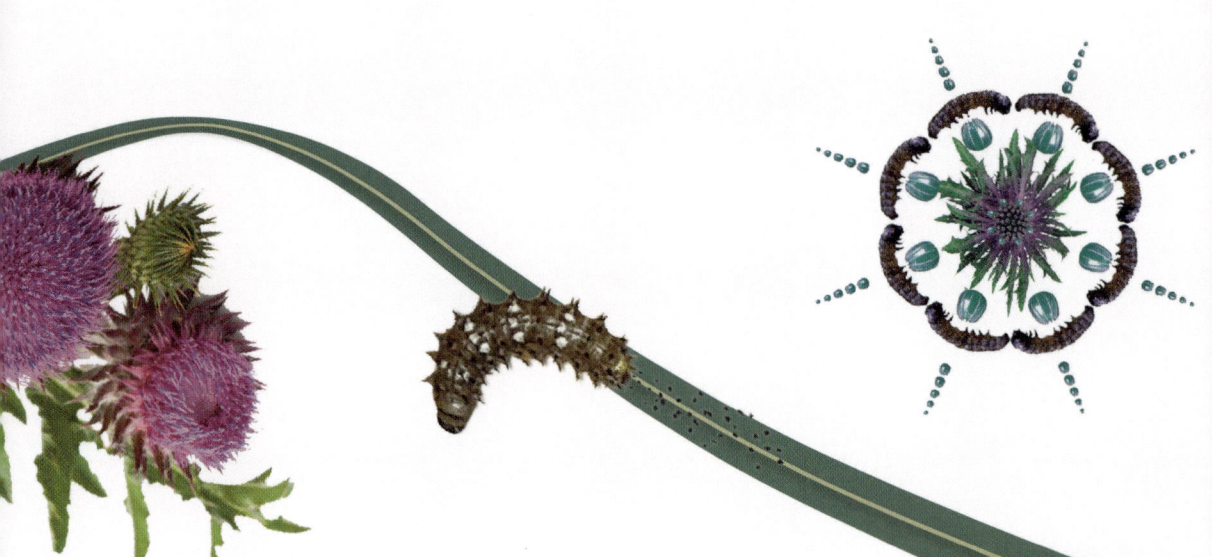

It was early April when Amatxi's Butterfly Taskforce was assembled. My siblings and I had less than a month to move from crawling to flying. We started out as wiggly larvae, so it was hard to imagine that we would ever fly.

But flying high and far is what we *Vanessas* are meant to do.

Altogether, I should say that my siblings and I had a pretty good childhood in Amsterdam. We ate, and we ate, and we ate; and we pooped, and we pooped, and we pooped, as caterpillars do.

I grew so quickly that every few days I had to shed my old skin, which at times became so tight that I could no longer grow unless I got rid of it. Human beings call this process *moulting* and it is no easy task, although it certainly feels liberating once you've done it.

It is liberating to shed old skin

Let me share a trick that we caterpillars use for getting rid of old skin, in case it might come in useful when you need to take off some tight clothes. First, we inflate our bodies until the pressure breaks open the skin of our back, as if it were a dorsal zipper. Then we wiggle out of the split skin and continue to eat and poop.

We have a bit of a wiggly advantage over you, since our cylindrical bodies are made of more than 4,000 muscles. You only have around 600, so wiggling out of tight clothes might be a bit harder. But the trick is a good one, in case you would like to give it a try.

Find a safe spot to transform

My childhood lasted about two weeks, during which I moulted five times. Altogether I went through five different caterpillar stages, which humans call *instars*, and shed five different skins.

My appearance changed a bit at each stage, and I was about ten days old when I entered my fifth and final instar. At this stage I was already preparing myself for something very special. I had developed *imaginal discs* inside my caterpillar body which, among other things, would serve to develop and attach the wings I was hoping to grow during adolescence.

Frankly, I was done with being a caterpillar and was ready for a radical transformation.

All of a sudden, I lost my insatiable appetite and all I cared about was finding a quiet spot where no one would bother me. Puberty was about to set in.

I was getting ready to form what you humans call a *pupa*, also known as a *chrysalis* if you are a butterfly. Since I would not be able to escape from pesky predators while pupating, I needed to find a safe place to transform, well protected from nasty creatures and from dehydration.

I found the perfect spot beneath a twig, where I spun a little patch of silk to help me hang upside down. Just that little patch was enough to secure my rear end to a twig, and I hung there, forming a J-shape with my body.

Everything in my caterpillar world was turning upside down. I made myself more compact, and moulted for the fifth and final time. The fresh skin of my new chrysalis quickly hardened to form a shield that would protect me through the next stage of my transformation.

3
The Imaginal Soup

One day, while I was still in my chrysalis, my siblings and I were carefully wrapped in soft paper and placed safely inside a box. We were carried to the airport and embarked on our very first flight, a 2,000-km journey from Amsterdam to Málaga as stowaways on a plane.

'Mission Amatxi' had begun.

Over the previous week, my entire caterpillar body had dissolved into mush. I was in the process of re-engineering my new eyes; a pair of high-tech antennae; three pairs of legs; and, most magical of all, two pairs of colourful wings to make me fly.

When we left home, my siblings and I had been in our chrysalis for between seven and nine days and we were almost ready to see the world with our butterfly eyes.

See the world with new eyes

Our rite of passage to adulthood, also known as *metamorphosis*, is not exclusive to butterflies. Many insects, and even amphibians such as toads and frogs, also experience it.

But I think it's fair to say that we butterflies undergo the most colourful of all transformations.

Beneath the stillness of our chrysalis, a perfect storm takes place – one that completely changes the way we sense and relate to the world. And one that will grant us the freedom to fly from flower to flower, tasting the sweetness of many nectars.

Trigger your imagination

As my chrysalis formed, dormant parts of my DNA became activated, dictating the release of some magical enzymes that scientists call *caspases*. These help to break down our caterpillar tissues into their basic components, such as molecules and simpler particles.

This might come as a surprise, but... did you know that you have caspases too? They are always working for you, helping you to get rid of old cells to make room for new ones. Without the work of your caspases, cellular debris would accumulate, which in turn could lead to the development of tumours. So, we should all be grateful to our caspases, don't you think?

During metamorphosis, the caspases of butterflies have to work extra hard. In the core of our chrysalis they keep busy brewing a greenish, chunky broth known as the *imaginal soup*.

Imaginal because of the *imago* or *adult* butterfly that will emerge from it. And, perhaps, because what I am about to tell you will likely trigger your *imagination*.

While my imaginal soup was boiling with activity,
a new kind of cells, the imaginal cells, appeared in the stew.
These new kids on the broth arrived with a wild dream.

They believed they could fly.

Not comfortable with change, my old caterpillar cells
fought against the newcomers and their ridiculous idea.
But more and more imaginal cells kept appearing
in my imaginal soup, vibrating together with the
same wild thought, and combining with each other
to gain strength.

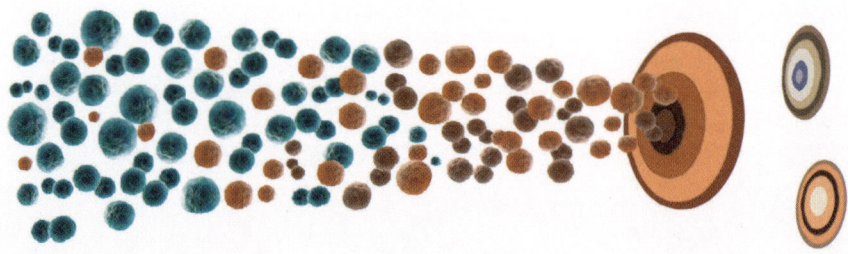

My old caterpillar establishment was rapidly losing
ground. Strings of new imaginal cells were clumping
together and taking raw materials to my imaginal discs.
Each imaginal disc contained the precise information
required to build a new body part: two new eyes, two
antennae, six legs and four new wings.

Imagine building a whole new you

Same DNA, different you

Altogether it took me ten days to build my butterfly body. Not bad at all, really, if you think that I had to revamp most of my nervous system so that it could orchestrate all my new functions, such as flying and sipping nectar.

I had to build a whole digestive system for my new liquid diet, along with an excretory system and a refurbished respiratory system.

By the time I was ready to emerge from my chrysalis,

I had recycled almost all my caterpillar components. The only waste product from my transformation was a drop of red fluid that humans call *meconium*.

I was still me. I still had the same DNA as when I hatched from my egg, the exact same building blocks I had as a caterpillar, but I had reincarnated into a fundamentally different creature.

Transformation is truly magical, isn't it?

4
A Fateful Accident

Still in our chrysalises, we landed in Málaga on a warm spring night. We were taken straight to Amatxi's home and placed carefully inside a netted space full of flowers and butterfly treats. The family called this 'Butterfly Headquarters', where we would stay for a couple of days until the arrival of Mother's Day – the time of Amatxi's special celebration, when we would all be released in her honour.

The morning after our arrival, some of my siblings and I were ready to come out from our chrysalises in the final act of metamorphosis, which humans call *eclosion*.

Emerging out of the tight quarters of a pupa is no easy task. I had to push and push with my legs, my antennae and my new long mouth until I managed to crack the chrysalis open.

Walking out of the paper-thin skin of my chrysalis, I looked for a place to hang from and stretch my wings. They were folded up like crunched sheets of paper. I really needed a good stretch.

Since I was no longer a herbivore endlessly munching thistles, I had swapped my previous chewing jaws for a *proboscis* – a long, thin tube adapted to sip sweet nectar out of lovely, colourful flowers. It's like a flexible straw that we keep coiled between our lips, the *labial palpi*, when not in use.

When we first emerge from the chrysalis, our proboscis is split into two parts along its whole length. We need to fuse these together into a single straw while we stretch and harden our wings. It took me a bit more than an hour to stretch my wings and fuse my proboscis.

A cool trick, which you might not know about, is that we can split our proboscis open again any time it gets clogged. That way, we keep it nice and clean.

Take time to stretch your wings

See the world through different eyes

Minding my own business inside Butterfly Headquarters, I perched and watched the world with my new eyes. These were so different from my caterpillar eyes, with which I could detect only light and dark. Now, I could look at the world with a pair of simple eyes adapted to see colour, as well as a pair of compound eyes that each contained 17,000 microscopic lenses. All the better to see you with, my dear!

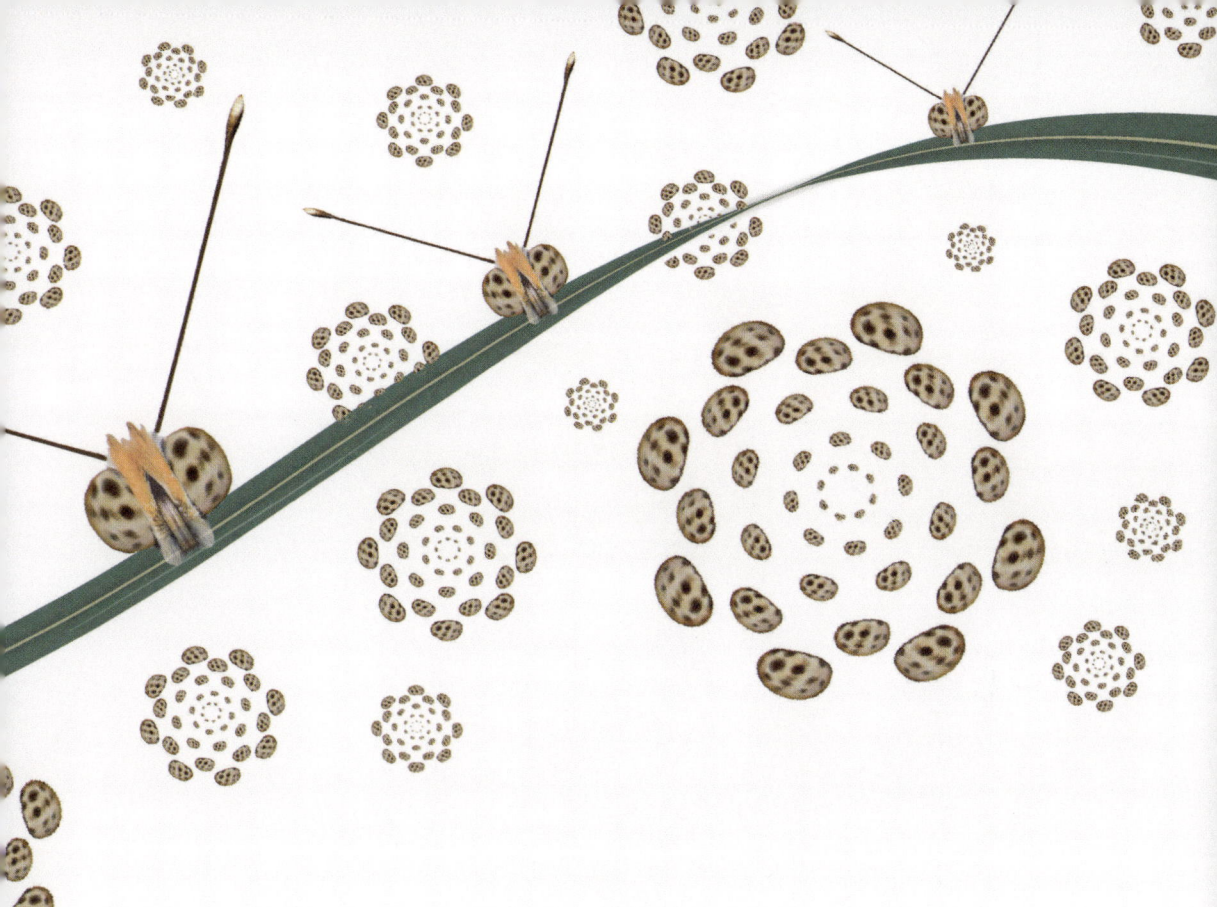

Working together, my two kinds of eyes, simple and compound, allowed me to see from as close as 1 cm to as far away as 200 m. It was amazing to look at the world this way.

I was checking out all the colourful flowers that surrounded me, when suddenly a humungous human finger appeared in my path. Feeling curious, I climbed on top of it.

And that single decision completely changed the course of my life...

Some flights lead to falls

I was carried on that giant finger from Butterfly Headquarters to the dining table, where the family had been eating lunch. Next, I was ceremoniously placed on Amatxi's finger. I had the honour to be 'The One' – the first butterfly to greet Amatxi!

In awe, she was admiring my colours when, all of a sudden... Oh, no! Amatxi's hand moved too quickly. I lost my balance, tried to take my maiden flight, but instead plunged straight into a bowl of leftover soup, right wing first.

Panic spread around the table.

Wet and scared, I was afraid that this would be the end for me. But the humans quickly fished me out of the lukewarm soup and dried me out with loving care. Although I had lost part of my right wing and the tip of my right antenna, they decided not to feed me to the lizards in the garden.

The family named me La Reme (*Reh-meh*), which in Spanish is short for the female name Remedios, meaning *remedies* – since I would be needing some of those to survive in the world as a flightless butterfly.

II

My Life as a Flightless Butterfly

5
Mission Amatxi

Mother's Day arrived two days after my accident. All my siblings had emerged from their chrysalises and were getting ready for the moment they would leave Butterfly Headquarters. This would begin their next adventure – a long, long flight known as the Great Painted Lady Migration. Dressed in orange, the many butterflies looked fantastic, sparks of life ready to conquer the skies.

Two pieces of technology – wings and antennae – would be essential for the long voyage ahead. Without part of a wing and with a shortened antenna, I stood no chance of joining my sibling butterflies on their journey. But my will to survive and reproduce was intact. I still hoped that some day, somehow, I would find a mate and pass my genes on to the next generation.

I wasn't going to lose hope.

Never lose hope

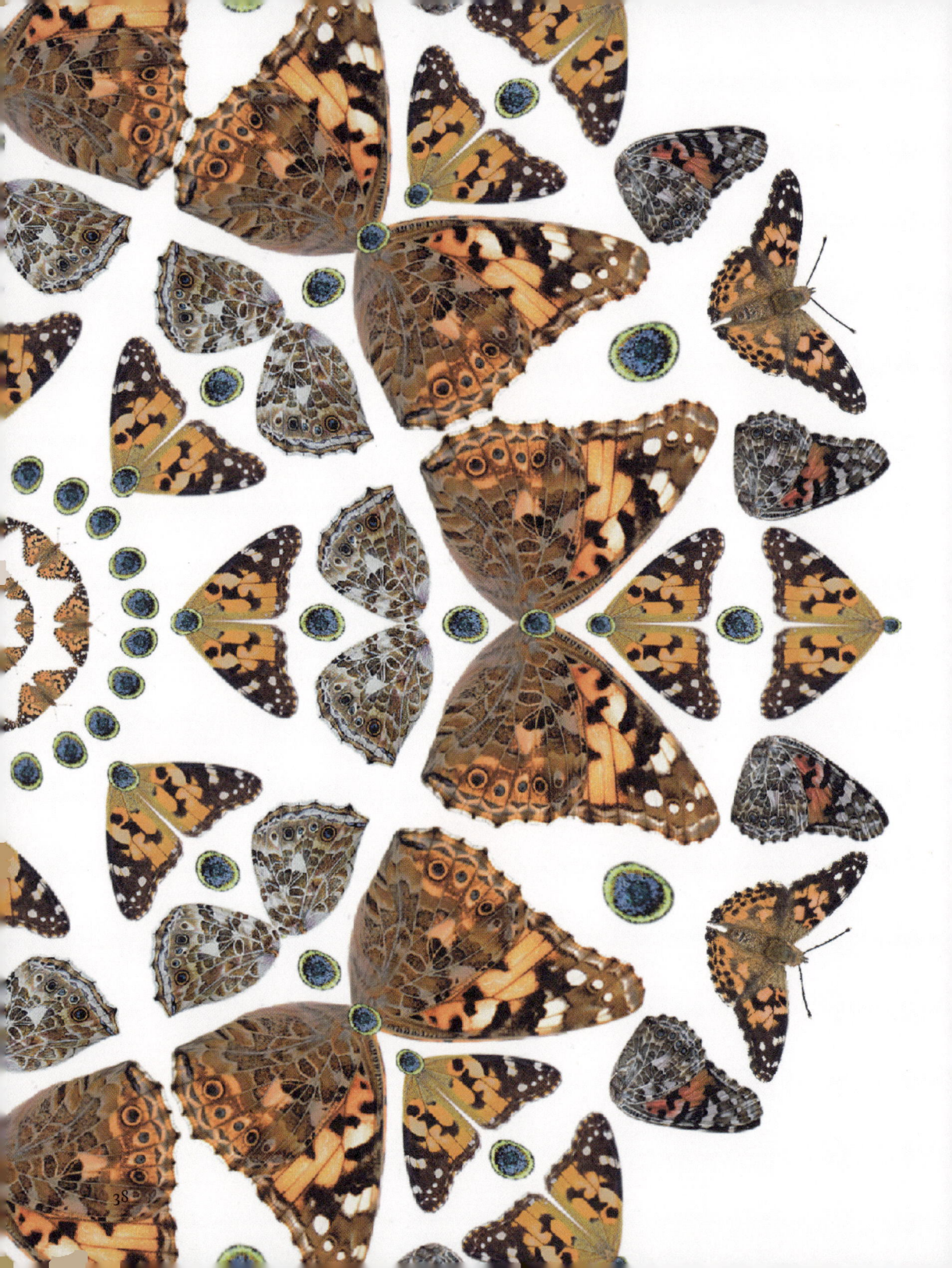

My antennae could still detect the alluring perfume of potential partners and my wings were perfectly capable of keeping me warm. This was crucial since we butterflies cannot produce our own body heat and rely on our wings to warm us up.

Wings can lift us into the skies but also work as highly efficient solar panels. They are made of tiny scales arranged like tiles on a rooftop, each with an intricate design that ensures the maximum absorption of solar energy. Wing scales have inspired humans to develop some of the most innovative designs in solar cell technology. Yet our wing structure, millions of years in the making, still outperforms any human engineering.

Our wings help us to communicate, and the unique wing design of each species allows us to identify potential mates. The patterns also work as camouflage to hide us from predators, especially when our wings are folded.

Some butterfly species have even evolved wing patterns that confuse predators in a fascinating, natural phenomenon that humans call *mimicry*. Wing designs mimic the appearance of dangerous creatures – like the eyes of an owl or the colours of a poisonous animal – and predators think twice before attacking them.

Camouflage communicates that there is nothing to see

We Painted Ladies have our personal take on mimicry. If you look at us from above, you will see that the core of our body, together with the brown parts of our wings, resembles a hairy spider. Not everyone is fooled by this trick, but at times it helps.

You might be surprised to know that some butterflies and most moths depend on their wings for hearing. Their ears are located on their 'armpits' – or *wingpits*, I should say!

Sometimes it is safer to look like a spider

To grant us a powerful sense of hearing, our wings work in concert with our antennae, which are excellent at picking up vibrations. Like our wings, our antennae are multifunctional. As you will see, they are an essential tool for guiding us as we travel around the world.

Our antennae help us to navigate

During metamorphosis our antennae are fitted with a sophisticated navigation system, which is crucial to our epic migration. Each hair-thin antenna is equipped with two types of compass and a wide variety of built-in sensors.

Our *sun compass* allows us to read the season and time of day from the position of the sun. Meanwhile our *magnetic compass* helps us to establish direction.

At the base of our antennae lies the *Johnston's organ*, a special piece of equipment that draws information from other parts of the antennae to assist with balance and orientation during flight.

The Johnston's organ also contains temperature and wind sensors – important both for flying and for identifying the wingbeats of other butterflies.

Besides working as a GPS, a radar, a clock and a hearing device, our antennae also work as noses. They are lined with special receptors that allow us to smell the sweet nectar of flowers and detect the enticing pheromones of potential mates.

You could say that we butterflies smell with our antennae. And, believe it or not, we taste with our feet, but I'll tell you more about that later...

After the accident, I could still use my antennae and my wings for many things, but instead of flying I had to walk.

I had no choice but to adapt to my new condition and overcome my fear of human fingers. I did so successfully, since finger-rides were the fastest way for me to travel. In fact, I started looking forward to being picked up and would walk up to fingers whenever they happened to show up.

We all need a helping hand at times

Riding on fingers is how I got to know Amatxi and her clan better than any of the other butterflies. The family learned how to handle me, I began to trust them, and so we adopted each other.

Since Amatxi was no longer able to walk on her own, she had to be taken to places in a wheelchair. I had to ride on fingers, so we both depended on the kindness of others to get from here to there.

On the afternoon of Mother's Day, the family finished their lunch fashionably late, as Spanish families tend to do. Then someone announced that the time had arrived for the Freedom Ceremony, when all the Painted Ladies would be released into the sky.

Along with the rest of the butterflies, Amatxi and I were taken out into the garden. There, the family gathered to express their gratitude to their beloved matriarch and to us butterflies for bringing so much goodness to the world.

Each of Amatxi's children and grandchildren held one of my siblings in their cupped hands, keeping them close to their hearts. Then, they gently opened their hands and wished farewell to my siblings while blowing a wish to the skies.

All the butterflies flew north, as we Painted Ladies always do in spring. Our two compasses work together to point us in the right direction – south in the autumn, and north in the spring.

Even though I couldn't fly along with my siblings, I was there, wishing them success for the long journey ahead. They would have to cut across all of Spain and continue on to northern Europe. That is one of the main routes of the Great Painted Lady Migration, which is the largest recorded migration of any butterfly species.

Express gratitude and blow a wish to the skies

6
The Gregarious Globetrotters

We Painted Ladies are the most cosmopolitan of the world's 20,000 butterfly species. That means you can find us pretty much anywhere on the planet. Humans consider us gregarious, because we love to hang out together in groups of hundreds of thousands, like butterfly raves.

We are true globetrotters. In fact, our migration is one of the furthest in the entire insect world. Our majestic relatives, the Monarch butterflies, also migrate over very long distances. They travel between Canada and Mexico, where they overwinter. But we Painted Ladies try to avoid winters altogether and are always on the move, tireless nomads of the wind.

In European summers you can find us breeding in very large groups in Norway. Actually, we breed in many places along the way, but we love the Norwegian meadows in the summertime.

Some Painted Ladies have flown as far as Iceland and the Arctic Circle, but usually we don't venture that far.

We are true globetrotters

In late August, our sun compass tells us that the seasons are changing, and it's time to head to Africa. We promptly start flying south on our next migratory odyssey.

Like all good navigators, we know how to read the winds that we depend on for our journey. Whether flying over mountains, sea or desert, we cross each massive landscape in one go, always catching the best air currents.

To get from northern Europe to tropical Africa we must fly over the freezing Alps, across the vast Mediterranean Sea and through the scorching Sahara Desert.

You can imagine that crossing the Alps is no small feat for a cold-blooded creature. We must fly at very high altitudes and in extremely cold temperatures. But can you imagine the equivalent journey for our Asian relatives? Twice a year, in their annual migration cycle, they must fly over nothing less than the peaks of the Himalayas.

You can look fragile and still conquer mountains

Painted Ladies might look fragile, but don't be mistaken, we are tough cookies. We have been seen flying at altitudes of 1.6 km above sea level, and at speeds of 50 kmph when blown by favourable winds. Sometimes we travel for over 4,000 km without a single pit-stop.

Like our lovely cousins the Monarchs, our full migration is an intergenerational process. During our spring migration, it takes at least two or three generations of butterflies to complete the journey from south to north. Therefore the Vanessas arriving in Norway in summertime are the granddaughters, and even great-granddaughters, of the butterflies that left Africa in early spring. However, in our autumn migration some Vanessas complete the journey from north to south in a single generation.

Altogether, we cover a staggering 14,000 km during our annual migration cycle, flying 7,000 km in each direction. And we fly together in groups of hundreds of thousands of migrating Vanessas.

When the time is right to come back down to earth, we land, we eat, we flirt, we flutter, we breed and – if we are still feeling strong enough – we continue on our journey. If we lay eggs, it's in our butterfly nature to die shortly after.

Our offspring continue the journey we started

Male butterflies tend to breed with many females, provided they get lucky, but they die shortly after they have used up all their fertilizing sperm.

One way or another, our offspring always continue the journey, north or south depending on the season.

If we don't get to mate during our lifetime, we will miss passing our genes on to the next generation, but we can potentially live up to a whole year. That's not bad in butterfly terms.

Amatxi had been around for more than 92 years – which is not bad in human terms.

She had led a cosmopolitan and gregarious life, travelling far and wide like a true Painted Lady. In Spain, where she grew up, she migrated each summer to the north, always returning to her wintering grounds in Madrid. But one day, while still fresh out of her chrysalis, she had found herself a lovely mate. They spread their wings together and flew non-stop to Puerto Rico without looking back.

All had been going well for Amatxi and her growing family on the Caribbean island when, just like me, an unfortunate accident changed the course of her life. Her lovely mate suffered a head injury and, after many trials and tribulations and hopes that he would recover, she had to accept that this would never happen.

She travelled far and wide like a true Painted Lady

Just as my accident changed my world, Amatxi's world had been turned upside down. She had six little offspring to raise – aged six months to ten years – and no time to think about anything else. She decided to fly back to Spain with her brood and her injured mate and start all over again.

And, so, she did.

I met Amatxi 54 years after she migrated from Puerto Rico back to her hometown in Spain. She had successfully raised her brood of six children and passed her genes further on to six grandchildren. She had lived kindly and generously. She had loved and was loved and had no regrets. But she was tired. The day I met her she was very weak, but having all the flutter of her family and the butterflies around her brought back her luminous spark.

Mission Amatxi had worked its magic.

After the Freedom Ceremony, I was the 'One and Only' butterfly left behind to continue with Mission Amatxi. I wasn't lonely, since I quickly became part of the family. Like us *Vanessas*, these humans were a gregarious and cosmopolitan bunch, and some were even globetrotters. I felt at ease around them.

Work your magic

Despite what had happened during our first encounter, I enjoyed hanging out with Amatxi. She was sweet and gentle, so easy to be with. Her children and grandchildren loved to sit around her, asking for stories. They would fly to faraway places in their imagination, while learning all about their ancestors. That's what they did after the Freedom Ceremony, and how I ended up learning so much about Amatxi's life.

As always happened when the clan gathered, they began telling stories, playing music, singing songs, and sharing laughter and joy. Everyone was cherishing being together, well aware that this would be the last celebration with their beloved ma and grandma.

Mission Amatxi had been a success. My siblings and I had brought joy to the hearts of Amatxi and her family and we could rest at ease.

Now her clan had to return home, myself included. I headed in the same direction as the other Painted Ladies, but took a very different kind of flight. Just as I had when I was in my chrysalis, I flew as a stowaway in a plane, carried carefully by Amatxi's youngest daughter, Trilin. Luckily, no one noticed a disabled butterfly going through airport security. After an effortless flight I landed in Amsterdam, back in my childhood home.

There is more than one way to get back home

7
Never Give Up

I arrived back in my birth home on a sunny afternoon, and was assigned a cosy corner in what Trilin called her Butterfly Palace. This is where she had first bred us, and it was teeming with *Vanessas* now. The last time I saw these younger siblings they had been caterpillars, too young to have been selected for Mission Amatxi.

They had stayed behind, and now they were full-blown adults having a wild party.

Two dozen brand-new butterflies were sunbathing, fluttering about, sipping orange juice and mating. In fact, some were mating and sipping orange juice at the same time! I was a bit shy, but looking forward to joining the party.

It would be hard for me to attract a partner, since part of our courtship ritual involves flying. But I could still produce a lovely pheromone fragrance, so I had hopes.

Between flowering plants, perfumed pheromones and all the fluttering *Vanessas*, I felt I had reached butterfly paradise. All of a sudden I had new friends, juicy flowers and the warm rays of the sun. Who could ask for more?

Even though I couldn't fly, my life at the Butterfly Palace wasn't at all boring. I received lots of visits in my cosy cove and every day I was taken out for sunny walks on the roof garden.

I was the 'One and Only' butterfly allowed to walk outside the palace, and I loved it. It was one of the perks of being different from the rest.

One afternoon, upon returning from one of my daily excursions, *he* came to my corner for a visit.

I had already spotted him flying up and down in a beautiful courtship dance. His fragrance was irresistible, the best I had ever smelled with my single antenna.

By then, I had almost given up on finding a mate. But there he was, for real, asking me for permission to fertilize my eggs.

He liked me just as I was,
without making any fuss
about my broken wing.

I was touched and flattered
and couldn't help but say,
'Let's do it!'

We mated for hours and hours
into the night, as butterflies
tend to do, locking the ends
of our abdomens so he could
pass his essence into me and
fertilize my eggs.

He liked me just as I was

We lay eggs only in the right plants

Butterfly eggs are tasty and nutritious, and nasty bugs love to eat them. We lay hundreds and hundreds of them – up to 1,300! – although less than half a dozen of them will make it to adulthood.

When we lay our eggs, it's crucial that we select the right plant, which humans call the *host plant*. Failing to lay eggs on the right one would be fatal for our babies.

Painted Ladies are less picky than most butterflies about their host plant. We lay eggs in several different kinds, but thistles are our absolute favourite.

We detect our host plant by tasting it with our feet, where we have chemical receptors that function like your taste buds. What's really special about tasting with our feet is that it allows us to detect if something is poisonous before ingesting it through our proboscis.

After the wild night with my mate, I was soon busy tasting plants and laying eggs, placing each one on the best areas to protect them from predators and dehydration.

Because of my condition, I was a bit slow and it took me two long days to place hundreds of eggs safely on several plants. In the meantime, my mate and the father of my future offspring was set free. I saw him flutter away along with the rest of this Painted Lady cohort, bright and beautiful against the blue sky.

Once again, I was the 'One and Only' left behind, wondering how it would feel to fly.

Not having other butterflies to play with, I ended up spending more time playing with my human friend, Trilin, as her family called her.

She had been taking care of my siblings and me from the very beginning, making sure we had a good environment to do our butterfly things and sneaking us on as stowaways when we had to take a plane ride.

I watched my mate fly away into the sky

If you open and close the book quickly,

you can MAKE ME FLY!

Make the most of your time together

After my friends flew away,
Trilin and I spent most of our
time together. I 'flew' around the
house perched on her finger. I also
earned my nectar as a model – posing for
Trilin as she learned to paint Painted Ladies.

As I had laid all my eggs, Trilin knew that I was
approaching the end of my life, so she wanted to
make the most of the time we had left together.

She was also struggling with the idea of losing her
beloved Amatxi. So I, in turn, tried to keep
her company and cheer her up.

III
Lessons from Nature

8
The Eternal Cycle

I died in the quiet of the night, the day after I laid my last egg.

It had been over five weeks since I was an egg myself, and two weeks since I emerged from my chrysalis. My life cycle had been fulfilled and my tired body was ready for recycling.

Amatxi's body was also weary. It had been 4,823 weeks (over 92 human years) since she was born from her mother's womb, and her life cycle too was coming to an end.

We had both lived full and eventful lives and had managed to pass on our unique genetic code to the next generation.

We had enjoyed ourselves. We had brought joy to others. The time had come to say goodbye.

We had brought joy to others

Early in the morning, Trilin found my lifeless body in the Butterfly Palace.

Sad, but accepting the ever-changing reality of life, she wanted to make sure that I would return to nature; back to the eternal cycle of generation, degeneration and regeneration of which we are all a part.

To wish me farewell, Trilin took me to the rooftop garden where I used to take my daily walks. With gratitude, she placed my body on top of a flower in a visible place.

I flew

A little while later, a blackbird spotted me.
He swooped down, picked me up and...
carried in his saffron beak...

I FLEW.

In the Butterfly Palace, the next generation of *Vanessas*, including my own children, had already transformed into colourful adults. A select group was recruited for the next mission, to accompany Amatxi on her final journey.

The new Butterfly Taskforce, assembled in Amsterdam, followed in our wingbeats and flew to Málaga as stowaways in a plane. Upon arrival, they were placed near Amatxi's bedside, to keep her company until the end and beyond.

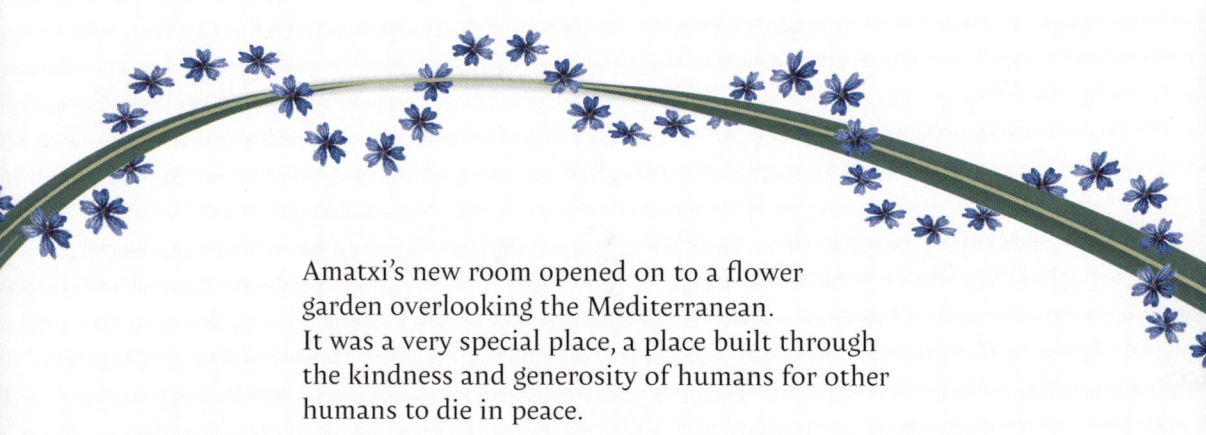

Amatxi's new room opened on to a flower garden overlooking the Mediterranean.
It was a very special place, a place built through the kindness and generosity of humans for other humans to die in peace.

Throughout the night, and all through the following day, Amatxi's life force kept fading away like the glow of a candle. She took her last sweet breath in the early afternoon, surrounded by her children and grandchildren.

A sense of deep peace and profound sadness settled over the room.

No longer would the family be able to hold her warm hands. Yet, they all knew that part of Amatxi, both through her memory and her DNA, would remain alive inside each one of them.

And they found comfort in that thought.

To soothe their grieving hearts, the family sat around Amatxi's body, and each shared a word about her character that inspired them:

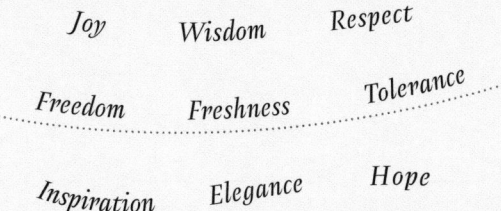

Joy *Wisdom* *Respect*

Freedom *Freshness* *Tolerance*

Inspiration *Elegance* *Hope*

All those attributes, and many more, were what their matriarch meant to them.

The time had arrived for the next generation of *Vanessas* to accompany Amatxi on the journey to her next stage. Placed over her silent heart, the Painted Ladies waited patiently.

Suddenly, in the blink of an eye, they all flew away, out into the garden and high into the clear, blue sky. The family wondered where they would go.

Wondering is something humans like
to do much more than butterflies, I think.

Have you ever wondered where your basic
elements have been since the beginning of
the beginning?

Every single particle that forms YOU has been
around since the Big Bang. That means that for
the past 13.8 billion years your constituent particles
have been recycled by nature countless times and
into wildly differing forms.

Have you ever wondered what new forms some
of your basic components will take in the future?

It is hard to imagine, but I certainly hope that some
of my components will meet again with some of
Amatxi's – some time, somewhere, in a joyful and
playful way.

Your particles have been around since the Big Bang

9
Looking Back

The day before I died, Trilin had asked me if I would like to share any special advice with her, just between butterfly and human.

I thought of all the different stages I had gone through in life – egg, caterpillar, chrysalis and imago – and pondered deeply on her question.

The following lessons are those I shared with her, based on personal experience. I'd like to share them with you, too, in case you find them helpful in embracing the changes in your own life...

Moult frequently

Moulting is hard, but necessary for you to grow. I told Trilin she should pause often to reflect on whether she was engaging in any behaviour that she needed to shed: something that could be hampering her own joy. She was raising an adolescent son and thought of a few personal behaviours she could dispose of in order to have more fun every day.

Shedding useless skin can be necessary to free yourself from your own shackles. My advice, from a butterfly perspective, is that you pause regularly to think about which personal thoughts, attitudes or behaviours are not helping you in any way. Then, just like a caterpillar, do anything in your power to shed that old skin so you can continue to grow nicely in life.

Like a caterpillar, shed that old skin

Choose your own wings

Imagine how to give wings to your dreams

Imagination is the first door to exploration, learning and discovering. Turn on your imaginal cells and dream up your world, just as butterflies do when we move from crawling to flying.

Trilin wanted to be able to create her own artwork, so she took my advice and pictured herself flying with ease like some artists do. She considered what she would need to do to move herself from crawling to flying, and imagined ways to get there.

My advice to you is the same advice I offered to my human friend. Think about the aspect of your life in which you would like to grow wings, choose your own wings, and imagine yourself already flying.

Wherever you wish to fly, use your imagination and never lose your sense of wonder.

Transform without fear

Change is the most constant aspect of life, you don't need to be afraid of it.

After imagining your new world and working out how to fly there, you must trust your own capacity to transform yourself. Trilin thought this was the most challenging lesson, since she really doubted she could actually create any interesting artwork. It is one thing to imagine it, quite another to actually do it.

I told her that profound transformation, such as the one needed to grow wings and antennae, requires you to reassemble some old components, while perhaps also shedding some limiting beliefs, behaviours or attitudes.

Trust in your own transformation

Deep change might seem impossible at first, but if you devote yourself to it you can transform yourself and your reality just by changing the way you interpret what's around you.

Don't be afraid to embrace your own transformation.

Collaborate with others

Be inspired by the work of the imaginal cells. They want to change their world and realize that only by working together in the form of imaginal discs can they become stronger and effective enough to debunk the old caterpillar establishment.

This piece of advice really appealed to Trilin. She was a firm believer in collaboration as the only way forward, particularly with her worries about how the natural world that she loved was being destroyed.

To go far, go together

Trilin was devoted to working with others and told me that she would look for ways to collaborate with humans and with butterflies to help us in our fight for survival. For my advice to you, I would like to use an old proverb:

'If you want to go fast, go alone; if you want to go far, go together.'

Keep this in mind and don't forget your own imaginal cells. They contain the blueprint of what you want to be in the future.

Celebrate your differences

Being disabled granted me a rather different and truly amazing life as a butterfly.

Everyone is unique and carries their own set of abilities and disabilities. Try not to get too hung up on your own imperfections, and celebrate the advantages you enjoy by being different.

Diversity is at the heart of the natural world. Without diversity, life can't thrive and evolve on our everchanging planet.

Appreciate and celebrate diversity

Celebrate your own differences and, just as importantly, celebrate what makes other people different.

The world is much more colourful and works much better when you appreciate diversity.

Observe nature

Nature has been figuring things out for billions of years, so there is a lot of wisdom accumulated there.

Everything in nature is in a constant state of change, and everything is interconnected in a wondrous web of life. You are an integral part of this web, so keep in mind that when you hurt nature you ultimately hurt yourself and your loved ones.

It's best to be kind to nature, to others and to yourself.

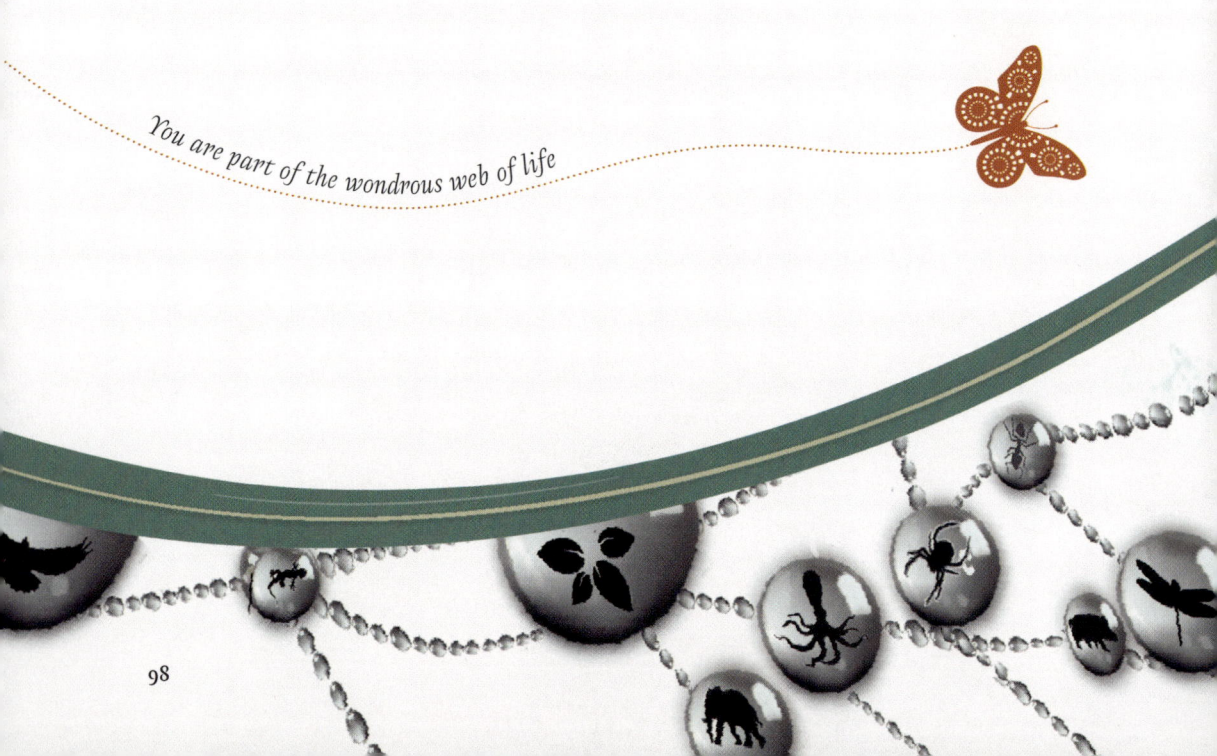

You are part of the wondrous web of life

IV
Regeneration

10

Attention, Here and Now

During my last evening I was very tired and quiet, but there was one more message I wanted to pass on to my human friend.

I told her that butterflies and humans needed to work together more than ever before. All butterflies are declining at an alarming rate, with two-thirds of the world's butterfly population disappearing over the past 50 years. We are being poisoned by industrial agriculture, and losing habitat to human overpopulation and overconsumption.

What humans don't seem to notice is that, without wild pollinators like us butterflies, your food crops will suffer greatly and your future will also be very challenging. For both our sakes, this needs urgent attention, here and now. We depend on each other, in our interconnected web of life, and we need to do something to change the reality we are all facing.

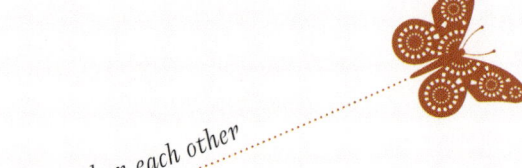

We depend on each other

Besides pollinating your crops, we butterflies help you to figure out if your environment is a healthy one or not.

As you humans put it, we are a great *indicator species*, because our presence indicates that your environment is healthy enough to harbour butterflies.

We are sensitive to change and to human poisons. So, if you see barely any butterflies around you it shows that your surroundings are not in good enough shape to harbour healthy life.

If you help us you will not only be protecting your own food production, you will also be supporting a significant part of the food chain.

Many animals, including birds, reptiles, mammals and insects, depend on butterflies for their survival. Our tiny eggs are very tasty, our caterpillars are packed with proteins, and as adult butterflies we are also delicious (except for the poisonous ones!).

Most of the hundreds and hundreds of eggs we lay are eaten. Whether as eggs, caterpillars, chrysalises or butterflies, we support the web of life, of which you are also a part.

By protecting butterflies, you help the pollination and productivity of your crops. You also help to protect biological diversity, which – in turn – is essential for your own wellbeing.

Butterflies have inhabited this planet for at least 55 million years and we would love to continue to hang around for a few million more. Evolutionarily speaking, human beings are the new kids on the block, having spent fewer than a quarter of a million years as a part of the web of life. But I bet you enjoy being a human being and would love your species to stay for a while longer on our shared planet. Wouldn't you?

The good news is that there are a number of things we can do to help each other.

11
Looking Forward

If we were able to collaborate like imaginal cells do, we could reconstruct a better world together.

This would be a world where butterflies, human beings and all sorts of wonderful creatures coexist with each other in a dynamic equilibrium. We have managed to do so until very recently in our interconnected web of life, and we can still recover that capacity.

We are all connected

So, what exactly can humans and butterflies do to help each other?

For our part, we butterflies can continue to do our work for free. We create beauty in the world. We pollinate your crops. We support the web of life. We help to measure the health of your environment.

And we bring joy to the hearts of many.

For your part, there are many things you can do for butterflies in return. You don't need to do them all, but you might find some that appeal to you...

Love us

They say that you can only love what you know. Now that you know butterflies a little better, you can love us a little bit more. If you wish to learn more about us, there are many sources of information about butterflies online. To get you started, I'll share some resources with you at the end.

Learn more about us

Don't poison us

Make wise food choices. Whenever possible, select food that has been grown without pesticides. Support the types of agriculture that are respectful of nature, such as organic and regenerative agriculture. You will be helping your own health, plus the health of your planet and the rest of its inhabitants.

Help your own health

Feed us

If you like flowers, plant some that are good for butterflies, with juicy nectar to give us energy. You could place these in pots on your windowsill or, if you are lucky enough to have a garden, make sure you have plants with flowers for butterflies and green leaves for caterpillars.

Natural gardens with an assortment of wildflowers and grasses are becoming more popular among humans, both in public areas and in private homes. These are very beautiful, require less water and maintenance, and provide food for pollinators.

If you live in the countryside and want to help us, promote the planting of hedges with a variety of aromatic plants and wildflowers. Make sure they are not sprayed with pesticides since that would kill us and defeat the purpose of supporting us. Hedges can be placed between crops or along fences and roads, or even planted in the form of small islands. We pollinators love those.

Having flowers and butterflies will offer the added value of creating beauty around you, which is always good medicine for the soul.

We all need a source of inspiration

Study us

Butterflies support researchers and educators in their task of explaining the intricate relationships that form the web of life.

We are a source of technological inspiration and have helped scientists to develop very useful technologies, such as highly advanced solar panels, strong fibres based on our silk, and even special pigments based on the structural colour of our wings.

Just as important, we are a source of poetic inspiration and a great metaphor for life, used by artists and philosophers. Whether in the sciences or the humanities, the study of butterflies is a regular source of reference.

Breed us

Not all butterflies are simple to breed, but we Painted Ladies are quite easy-going. We are often bred in schools to teach children about life cycles and metamorphosis. You can also breed us at home to help raise awareness about our worldwide decline and our role in the web of life.

If you do decide to breed butterflies, it is important to select species that are suitable for your environment. If you release the wrong species, you might end up creating a problem for the local butterflies, so you need to research that carefully.

Raise awareness

Monitor our migrations

Track us

There are apps available to help you identify butterflies and share your sightings with a global network. If you've fallen in love with us *Vanessas,* there is an app to help you monitor our migration and the migrations of other butterfly species. I'll give you some pointers at the end.

Make a difference

Protect us

In every country, there are butterfly-conservation associations with knowledgeable people who know how to help us. If you cannot devote time to us directly, you can always support the work of your local butterfly network and make an important difference in that way.

12
Stardust

Whether you are a butterfly or a human being, you are made of stardust that has been circulating in the cosmos for billions and billions of years.

Some of these particles are presently rearranged into a unique form that is alive here and now – that brilliant composition we know as YOU.

Be mindful that you are very privileged to be a part of the web of life. Be respectful of other living beings, and enjoy the many transformations of your own life adventure.

You are a brilliant composition

In Memoriam

Beatriz Gómez-Urrutia
'Amatxi'
1926–2019

Disappearing Butterflies

Worldwide butterfly diversity and abundance has declined by 70% between 1970 and 2020

2020

Further Flights

INSPIRATION

The author is donating her royalties from this book to support butterfly conservation projects.

Inspiration 4 Action
inspiration4action.com

This is the organization through which Astrid Vargas works with individuals, communities and organizations to help develop creative ways of nurturing nature and achieving regenerative change in landscapes and communities.

Regenerating Butterflies
inspiration4action.com/projects/regenerating-butterflies/

Astrid is part of a landscape restoration initiative in the high steppes of southern Spain. Local schools and municipalities take part in the 'Regenerating Butterflies' project, which promotes the planting of large hedgerows of aromatic plants arranged in the shape of a butterfly... to feed the butterflies. These regenerative sculptures serve as an educational resource, restore the soil, retain water, promote biodiversity and strengthen the bonds between people.

CONSERVATION

UK
butterfly-conservation.org

The world's largest research institute for butterflies and moths, Butterfly Conservation operates 36 nature reserves and is involved in over 70 major projects to conserve habitats.

Australasia
maba.org.au

Moths and Butterflies Australasia aims to encourage interest in the scientific study, research and conservation of moths and butterflies in the Australian region.

Europe
vlinderstichting.nl

Butterfly Conservation Europe is a network of partner organisations for the conservation of Lepidoptera and their habitats.

United States
naba.org/butterfly-counts/

The North American Butterfly Association has 23 chapters across 14 states, and works to create a world where butterflies thrive, for the benefit of nature and people.

OBSERVATION

Big Butterfly Count UK
bigbutterflycount.butterfly-conservation.org

Take part in the UK's annual Big Butterfly Count by spending 15 minutes in nature identifying and counting the relevant butterfly and moth species you spot, then submit your sightings to help monitor how butterflies are faring.

eBMS
butterfly-monitoring.net

The European Butterfly Monitoring Scheme is a pan-european citizen science initiative.

EXPLORATION

Use your phone to scan the QR code below and discover the Linktree for *On A Butterfly's Wing*. Here you can take flight from the book and explore a world of butterfly resources.

About the Author

Astrid Vargas (who is 'Trilin' in the story) is a conservation biologist with a track record in setting up, developing, leading and monitoring environmental restoration programmes. She has been a key figure in the recovery of three of the world's most endangered species: the Iberian lynx in Spain; the black-footed ferret in North America; and the golden-crowned sifaka in Madagascar.

Astrid was named by *El País*, the Spanish newspaper of record, as one of the Top 100 people in Ibero-America – men and women who have made a difference. She has earned multiple awards including, among others, a Lifetime Career Achievement Award from the Government of Andalusia, a Nature Conservation and Research Award from EDC, Spain, and a top 10 nomination in the global Future for Nature Awards, The Netherlands.

Astrid believes in the power of art and inspiration to transform society. She is the founder of Inspiration 4 Action, an initiative that inspires communities to bring collective creativity into ecosystem restoration.

inspiration4action.com

Gratitude

A mountain of gratitude goes to Amatxi, my mother. So much strength and gentleness combined in such a loving and lovable human being. Her wholehearted support empowered those around her to grow their own wings, and for that – and for so much more – I am forever grateful.

This book is a tribute to Amatxi and her clan: Beatriz, Marina, Javier, Laura, Nicky, Carlota, Rodrigo, Candela, Beatriz, Mario, Darío, Daphne, Virginia, Fernando, Ricky and Joe, with much love from me, Trilin, the youngest of Amatxi's six children.

Joe Zammit-Lucia, my husband, was an enormous source of support and inspiration throughout the writing of *On a Butterfly's Wing* and beyond. Thank you, Joe, for your big-hearted generosity. Mario, my son, was always game to discuss design matters, and provided great suggestions that influenced many illustrations. Thank you, *mi niño*, for your creative ideas and for thinking outside the box.

Jason Hook, Jenny Manstead, Luke Herriott and the Unipress team were a blessing to work with. I am truly grateful for their support, expertise and meaningful advice for the publication of this book. My deep gratitude also goes to Ricky Smith, Daniel Iglesias, Lucía Calzada, Jacinto Román, Guyonne Janns, Eloy Revilla, Miguel Delibes, Constantí Stefanescu, Nela Alvarez, Amatxi's Clan, and the many wonderful friends who took a look at the story and its illustrations and provided encouragement at all stages of its evolution.

The digital collages that illustrate this autobiography of La Reme are composed of photographs I took while rearing butterflies on the rooftop of my home in Amsterdam. I want to thank the butterflies that surrounded my world for the countless hours of joy, discovery and inspiration, with special gratitude to La Reme for being such a cherished companion. Some digital mandalas were inspired by the beautiful work of artists Christopher Marley and Damian Hirst. Maria Popova's literary blog, The Marginalian, has been a source of philosophical inspiration for years, influencing much of the self-reflection behind this book.

The stunning butterfly images showcased in the illustrations from Chapters 8 to 9 were generously donated by top-notch conservation biologists and dear friends Jacinto Román and Eloy Revilla. A very special acknowledgement goes to Constantí Stefanescu and his team at the Natural History Museum of Granollers in Catalunya, Spain, for their continuous efforts in uncovering remarkable findings on the migration patterns of the Painted Lady butterfly.

My heartfelt gratitude to Tony Tabatznik, who believed in the value of this story as an educational tool and helped its publication through the support of Bertha Foundation. And much gratitude to David Forbes-Nixon for all his support.

Finally, I would like to thank all those who for years and decades have devoted their time and energy to the study and conservation of butterflies – and continue to do so. Their work is priceless and a source of both knowledge and inspiration for us all.

First published in the UK in 2024 by
Riverside Press
an imprint of
UniPress Books Ltd
World's End Studios
London SW10 0RJ
United Kingdom

Text copyright © Astrid Vargas 2024
Images copyright © Astrid Vargas 2024
Copyright in the Work © Unipress Books Ltd 2024

The author has asserted her moral rights.

ISBN 978-1-7397988-5-7
E-book ISBN 978-1-7397988-6-4

All Rights Reserved. No part of this publication may be reproduced, stored in a retrieval system or transmitted in any form or by any means, without prior permission in writing from the publishers.

A catalogue record for this book is available from the British Library.

Publisher: Jason Hook
Design: Luke Herriott
Original Design: Daniel Iglesias & Mario Vargas
Illustrations: Astrid Vargas
Colour Reproduction: Les Hunt

Printed in China
10 9 8 7 6 5 4 3 2 1

unipressbooks.com

This book was printed on paper from sustainable sources.